# YOUR KNOWLEDGE HAS VALUE

- We will publish your bachelor's and
  master's thesis, essays and papers

- Your own eBook and book -
  sold worldwide in all relevant shops

- Earn money with each sale

Upload your text at www.GRIN.com
and publish for free

# The Environmental Physiology of Coffee. An Advanced Coffee Review

Tamene Haile Kitila

**Bibliographic information published by the German National Library:**

The German National Library lists this publication in the National Bibliography; detailed bibliographic data are available on the Internet at http://dnb.dnb.de.

ISBN: 9783346894106
This book is also available as an ebook.

WALLAGA UNIVERSITY SHAMBU CAMPUS

SCHOOL OF POST GRADUATE

ECOPHYSIOLOGY OF COFFEE

BY TAMENE HAIE KITILA


**FACULTY OF AGRICULTURE**

**DEPARTMENT OF HORTICULTURE**


**PROGRAM: MSC. IN HORTICULTURE**


**COURSE TITLE: ADVANCED COFFE PRODUCTION AND PROCESSING (HORT 5071)**


**JUNE 2023**

**SHAMBU, ETHIOPIA**

# ACKNOWLEDGMENT

I would first like to express my gratitude to the Almighty God for the success of my daily tasks and the completion of this review preparation. I want to take this opportunity to thank Dr. Abera Daba for helping me complete my paper and for directing me in doing so in order to improve both my writing abilities and my understanding of coffee physiology towards ecology. Furthermore, I want to thank Mr. Desalegn Worku for his advice and comments that were beneficial to me as I was writing my review, especially when it came to illustrating the structure of Wallaga University's paper writing styles. A special word of thanks goes out to the horticulture department for providing me with the materials I needed to complete this paper.

# TABLE OF CONTENTS

# ABSTRACT

Coffee is one of the most important global crops and provides a livelihood to millions of people living in world. Coffee species have been described as being highly sensitive to climate and soil factor as largely affects coffee physiology. Here, physiological responses of the coffee tree in the context of present and ongoing climate, including drought, temperature, light and shade and interactions between these factors is discussed. The physiological and agronomic performance of coffee at different altitudes is explained. Evidence is shown suggesting that warming, perse, may be less harmful to coffee suitability than previously estimated, at least under the conditions of an adequate water supply. Finally, the effect of both wet and dry wind, as well as humidity is included under this paper.

**Keywords**: Coffee spp., coffee, drought, elevated, light stress, photosynthesis

# 1. INTRODUCTION

Coffee (*Coffee arabica* L.) is widespread throughout the tropics with more than 70 species. All cultivated species originate from Africa. Coffee arabica, which accounts for 64% of global production, and Coffea canephora (var. Robusta), which accounts for 35%, are today's two most significant coffee species economically (ICO, 2020). Coffee is one of the most significant agricultural products on the global market; it is grown on a total of 10.3 million hectares globally and provides more than 25 million families with their sole source of income. The crop is grown and sold by more than 60 countries and is one of the most important cash crops in developing nations (Pohlan, H. A. J., and Janssens, 2010).

Brazil is the largest world's coffee producer, followed by Vietnam and Colombia. Coffee is the major export product of some countries such as Uganda, Burundi, Rwanda and Ethiopia. About 70% of the world crop is grown on smallholdings smaller than 10 ha, and hence it is often a family business that provides maintenance for over 25 million people worldwide (Davies et al., 2006),

According to Wrigley (1988), coffee (*Coffee arabica* L.) is grown in a variety of habitats that stetch 25° north and south of the equator.Within these habitats, numerous growth and floweri ng paterns are seen.Growth flushes and flowering are seen all year long in countries that are cl ose to the equator yet lack well defined seasons, as Colombia (Morales *et al.,*1951).Contrarily , there is only one growth and flowering season in Brazil and Ethiopia, both of which have uni que seasonal climates (Barros and Maestri, 1972; Clowes and Allison, 1982). The key to mechanical harvesting lies in flowering, which also causes issues with fruit ripening. There is no information on the environmental conditions that favor a brief, abundant blossoming. Coffee is grown all over the world within the confines of tropical and subtropical regions. Because of this, coffee is a tropical plant that grows best between the latitudes of 25°N and 25°S, but different species need highly particular environmental conditions for commercial cultivation (Coste, 1992; Wrigley, 1988).Because resource availability restricts plant growth and biomass allocation patterns, at the whole plant level, evaluation is necessary (Poorter and Sack, 2012).

The literature on how environmental factors affect the growth and flowering of coffee under field circumstances is unclear. Depending on the location, the regularity of the wet and dry seasons, temperature, radiation, and photoperiod may each be the dominant factor controlling the vegetative/reproductive cycle, which can extend over several months even in places with distinct seasons and pose significant challenges for effective mechanical harvesting (Gopal, 1974). Establishing the finest methods of leadership and creating coffee production systems benefit from knowledge of the ecophysiology of coffee.

The growth and development of the coffee plant can be affected by a change in rainfall, altitude (Cheserek and Gichimu, 2012), temperature, relative humidity, light, moisture, and soil nutrients (Steiman, 2013). The combined effect of these factors resulted in a change in the inherent characteristics and final beverage quality of coffee (Barbosa et al., 2012; ITC, 2011). The change in the beverage quality of coffee comes from the effect of important biochemicals that are affected by diverse environmental conditions (Sridevi and Giridhar, 2013). In addition, the lack of proper harvesting and processing methods are thought to be contributing factors to the poor quality of coffee (ITC, 2011). Ethiopian coffee is grown between 1500-1900 m above sea level. However, it is also possible to find coffee at lower (1000 m) and high (2500 m) elevations. The Southwestern areas have a diverse altitude range, lengthy rainy seasons, variable cropping systems, and the presence of intact natural habitats which are suitable for coffee production (Gole et al., 2015; Kufa, 2010; ). In the area, macro soil nutrients (P and K) and texture characteristics (silt and clay) were identified as positively correlated nutrients with better cup quality of forest coffees. Furthermore, higher levels of major soil elements like Mg, Mn, and Zn and pH were related to coffee aroma (Yadessa et al., 2009). These all-diverse features could give coffees of good quality specific origins and tastes (Laderach et al., 2011) that can attract international markets that depend on selecting specialty coffees (Belete et al., 2014).

While discussing as a review, this work does not go into great detail but rather concentrates on ecophysiological elements of coffee production. Sections dealing with temperature, light and shade, altitude, wind, humidity, water or rainfall, and soil type are included in this paper.

# 2. ENVIRONMENTAL PHYSIOLOGY OF COFFEE

Since the interaction of many factors, including temperature (minima, maxima, mean), rainfall amount and distribution, type of soil, and topography, play a significant role in determining coffee physiological performance, it is difficult to pinpoint precise environmental thresholds for coffee cultivation. Although it is a plant that prefers to be in the shade, coffee can often grow well without it and can produce more than coffee that is grown in the shade. In comparison to many other tropical tree crops, the species has a very low rate of net CO2 assimilation (usually between 4 and 11 mol m-2s-1 with the current atmospheric CO2 content and saturating light; Araujo *et al.*, 2008).

High irradiance exposure to coffee plants can have a negative effect on their performance by restricting photosynthesis, which reduces net carbon gain and plant development (Grades, 2007). The crop's leaves won't have enough reaction centers to store light energy and transform it into biochemical energy if the light intensity is too high. Because of this extreme photorespiration,                                                                                the majority of the carbohydrates in the coffee plants' reserves eventually run out, as a result,the tr eescould experience a severe dieback. Furthermore, in unshaded coffee orchards,problems incl uding excessive evapotranspiration, severe drought stress, the death of actively growing shoot sections, periodic leaf crinkling, frost damage, and subsequent production decline are frequent occurrences.Although this depends on competition for water and nutrients, pest and disease in cidence, coffee plants do demonstrate some ability to acclimatize to sunlight (Grades, 2007).

High temperatures frequently accompany climate circumstances where plants must adapt to ex posure to high irradiances, and one consequence may be compounded by the other.In contrast, coffee is routinely exposed to temperatures of 40°C or more due to the widespread practice of growing coffee in the open sun (Araujo *et al.*,2008),which is more than double the ideal temperature ra nge for Arabica coffee.As a result,warming of the leaves is a typical occurrence,which may ha ve disastrous repercussions on plant metabolism by upsetting cellular homoeostasis and decou pling crucial physiological processes. The end effect is not only slowed growth but also a lack of storage space, flower abortion, an acceleration of fruit development and ripening, and a resulting decrease in quality (Grades, 2007). Despite the fact that coffee is a shade-loving plant with a higher quantum utilization efficiency for photosynthesis, excessive shading or light interception by the top two to three canopy strata of different tree species would reduce crop growth and productivity because the plant uses a large portion of its photosynthetic resources for maintenance (TayeKufa, 2007). The ability of coffee plants to adapt to environmental changes is crucial for their growth, development, biomass accumulation, and productivity (Ramalho *et al.*, 2014). This ability is influenced by the speed and length of the exposure as well as the temperature range.

Both C. arabica and C. canephora exhibit physiological and metabolic effects of cold, similar to other tropical species. These effects include changes in net photosynthesis and in some sensitive areas of the photosynthetic apparatus (such as photosystem (PS) II), which are frequently accompanied by leaf tissue damage and shedding (Batista-Santos *et al.*, 2011). These effects were related to the cold induced oxidative stress, which was connected to an excess of excited oxygen and chlorophyll molecules (Fortunato *et al.*, 2010). Low temperatures can also harm the root system of the coffee plant (Queiroz *et al.*, 2000). However, certain coffee genotypes showed some meaningful cold acclimatization capacity. This was determined to be related to the ascorbate glutathione cycle enzymes and the photo protective xanthophyll cycle, as well as qualitative changes in the lipid matrix of chloroplast membranes (Scotti Campos *et al.*, 2014) that maintained an adequate flexibility to membrane-related functions, like thylakoid electron transport (Batista-Santos *et al.*, 2011).

## 2.1. Climatic Effect on Coffee Physiology

### 2.1.1. Temperature

The most significant climatic component that significantly affects the physiology of the coffee plant is temperature (Carr, 2001). If the temperature changes are not too severe, Arabica coffee can resist them (Taye Kufa, 2006). Although it may survive temperatures much lower or higher than these limits for brief periods, the ideal average temperature is between 15 and 24 °C (Taye Kufa, 2006). Physiological processes in daytime air can be affected by temperature increases exceeding 30°C. Furthermore, ongoing exposure to temperatures as high as 30 °C causes an overall reduction in tree health and promotes leaf loss (DaMatta and Ramalho, 2006). The highest growth rates are shown from September to May, when the lowest and maximum temperatures are greater than 17 °C and less than 34 °C, respectively, for the principal C. canephora genotypes (such as Conilon and Robusta), which are typically planted at latitudes between 18° and 22° south (Partelli *et al.*, 2013).According to Rodrigues et al. (2016), this species prefers an average temperature between 22 and 26 °C. When exposed to temperatures below those ranges, photosynthesis and growth are significantly reduced (Batista-Santos et al.,

2011). Because low temperatures are common at altitudes greater than 500 meters, C. canephora has been deemed to have limited ability to adapt to them. In contrast, C. arabica plants cultivated at latitudes between 16° and 20° south experience their fastest growth rates from September to March, when the minimum and maximum average temperatures are, respectively, greater than 15 °C and less than 30 °C (Ferreira *et al.,* 2013). Genotypes from this species are better suited to withstand the somewhat lower temperatures found above 500 meters since they may survive brief periods of temperatures as low as 4 °C (Ramalho *et al.,* 2014).

According to Fekru Meko (2005), temperatures above the optimum encourage rapid growth, early and excessive conception, early tiredness, dieback, and disease attack. In addition to blossom shedding and diminished fruit development, higher temperatures can also result in slow, stunted, and unprofitable growth as well as a large production of secondary and tertiary branches and the so-called "hot-and- cold" illness. According to Willson (1985), temperatures below 12°C impede growth and development, and above 24°C, net photosynthesis starts to decline and becomes insignificant at 34°C. Frost also damages both leaves and fruits. Altitude-modified climates are preferred by arabica coffee. Taye Kufa (2006) explains that physiological issues brought on by temperature drops can impact coffee plant growth and fruiting at higher altitudes. Clowes *et al.* (1989), Stated that temperature has a significant impact on coffee plants' vegetative growth and floral start. They discovered that vegetative development is best around temperatures of 23°/17°C. The report claimed that low temperatures would encourage consistent floral initiation, which in turn would encourage uniform flowering. Because floral initiation happens gradually along the branch, it reduces the length of time that vegetation is actively growing. Low temperatures would encourage the start of floral initiation as well.

Larcher, (2000) asserts that one of the most important environmental elements is temperature, which has a significant impact on all physiological processes in plants and regulates the levels of metabolic activity within cells. Any higher or lower differential in temperature could reduce the efficiency of any and all physiological activities and chemical interactions.

Different plants are more or less tolerant of severe temperatures, and each one has certain growth and reproductive restrictions. For instance, even though a plant's seed may germinate and grow at a certain temperature, the same temperature may not be ideal for the reproductive stage. Therefore, the plant won't be required to colonize that habitat if the temperature is suitable for only some stages of plant growth and development (Schulze *et al.,* 2005).

Due to stomatal and non stomatal variables, low temperatures have a negative impact on coffee photosynthesis and yield (Ramalho *et al.*, 2000b). Arabica coffee is frequently grown in shade in areas with somewhat high mean annual temperatures, as is the case in Central America (Maestri and Barros, 1977). This has historically been attributed to the fact that net photosynthesis would drastically decline above 24 °C and eventually reach zero at 34 °C (Nunes *et al.*, 1968). The internal $CO_2$ concentration of coffee leaves grew logarithmically up to 30°C and then more quickly to 35°C, according to Heath and Orchard (1957), indicating a thermal limit on photosynthesis.

Numerous studies have demonstrated how temperature affects the growth of coffee plants as well as how it affects flowering time and yield in general (Oliveira RR *et al.*, 2020; DaMatta FM *et al.*, 2019), with a negative warming scenario expected for the ensuing decades (Bunn C et al., 201).Because the 'optimum' temperature also relies on the species, phenology, and developmental stage of the plant, evaluating the effects of climate on coffee plants is much more difficult. The ideal yearly temperature for C. arabica cultivars typically falls between 18 and 23 °C, whereas the ideal temperature for C. canephora is between 22 and 26 °C (DaMatta FM *et al.*, 2019). Both species are being studied to determine the impacts of either progressive temperature increase or high heat stress. Minimal impact on photosynthetic-related parameters occurs in leaves when plants were exposed to temperatures up to 37 °C, but severe damage occurs at 42 °C (Martins MQ *et al.*, 2016)

However, this effect varies according to the tissue and species because, when exposed to 45 °C for 11.5 hours, C. arabica appears to develop aberrant reproductive structures (DaMatta FM *et al.*, 2019), whereas no such effects were noted for C. canephora. Accordingly, in C. arabica, high temperatures accompanied by severe water deficit levels at the start of the anthesis period result in pollen tube dehydration and floral atrophy, which causes flowers to open prematurely and abort, with the petals remaining small and stiff and taking on a starlet shape (Mayer JLS *et al.*, 2013).

Although the effects of temperature on coffee plant growth in terms of leaves and bean output are well known, the effects of temperature on flowering have not yet been quantified, and the associated molecular pathways have not yet been identified. Temperature is a significant factor in flowering control that is linked to circadian clock genes and photoperiodic pathways. It has been shown that constants are responsive to high and low temperatures to influence FT transcription and protein accumulation (Martins MQ *et al.*, 2016).

A temperature also has an impact on the coffee root's physiological processes and root growth. In fact, Barros et al. (1997) reported that transitory reduction in shoot growth during the active growing season appears to be primarily connected with high temperatures. Dropping temperatures would in large part induce the declining growth rates through the quiescent phase. As the beginning of the falling growth phase was also accompanied by a parallelism between stomatal conductance and shoot growth, the researchers made the hypothesis that photosynthesis might, to some extent, influence the growth of the coffee tree. However, lowering air temperature appears to be the cause of declines in both growth and photosynthesis, and these declines would simply run concurrently rather than being connected, according to Silva *et al.* (2004)

In general, coffee prefers a mean yearly temperature between 18 and 22 degrees. Fruit development and ripening are accelerated at temperatures above 23° C, and this might result in a loss of physical quality as well as beverage quality. A protracted dry season and high temperatures above 30°C during flowering may result in floral abortion.

## 2.1.2. Light

Even though light is a key element of the environment that promotes photosynthesis and eventually affects plant growth, both low and excessive levels of sunshine can have a negative impact on a plant's ability to function. Despite being thought of as a shade-dwelling species, commercial coffee species have been produced all over the world in a variety of lighting conditions, from intense sunlight (as in Brazil) to relatively deep shadow (as in some regions of Central America where C. arabica is farmed).According to Char bonnier, F. *et al.* (2017), coffee exhibits a noticeable phenotypic plasticity to adapt to changes in light availability at both the leaf and whole-plant levels.

In any event, under intense management conditions, coffee grown in full sunshine frequently outperforms coffee grown in shadow in suitable areas. This is most probable because elite coffee cultivars were chosen during test trials in Brazil that included substantial external inputs and full sunlight. As a result, shade has either been abandoned as a common cultural practice or significantly reduced in scope in many areas of the world (DaMatta, F. M., 2018).When the portion of energy needed for photosynthesis decreases (for instance, during drought circumstances), while the absorbed energy remains constant (or even grows), a surplus of light energy can develop in the photosynthetic system.

However, in many instances, an abundance of energy is connected to high sunshine, which can limit plant performance, mostly by aggravating oxidative stress. It is common for linear electron fluxes to be many times higher than those needed for the measured A when low A in coffee leaves is combined with high irradiance levels (Martins *et al.,* 2014) .However, even in leaves that are significantly exposed to direct sun radiation, photo inhibition and photo damage are rarely seen when coffee plants are grown in optimal climate circumstances. In fact, a variety of energy dissipating pathways provides coffee leaves with excellent protection against oxidative stress when they are fully exposed to sunshine. (S. Krish, *et al.,* 2013)

Photo inhibition is a phenomenon that is caused by excessive light energy, either alone or in combination with other environmental factors such as water scarcity and high temperatures. As a result, plants must have mechanisms in place to either release this excess energy or, alternatively, repair any potential damage caused by photo inhibition (Pearcy, 1998).Demmig-Adams and Adams (1992) demonstrated that down-regulating PS II in high-light settings prevented the photosynthetic reaction center from being over excited and that the xanthophyll cycle, which is found in thylakoid membranes, is involved in the disposal of extra energy. According to certain research, coffee can respond to excessive light (1,500 mol m$^{-2}$s$^{-1}$ over the course of 6 to 8 hours) by producing more xanthophylls and other defensive pigments in the leaves. This response is also correlated with nitrogen availability (Ramalho *et al.,* 1997). Less grana and thylakoids per chloroplast are found in plants growing in full sunshine, indicating that the cells have adapted to deal with the abundant light (Fahl *et al.,* 1994).

Coffee grows in shady or somewhat shaded areas in its native environment. Its reaction to light has led to it being historically regarded as a heliophobic plant that needs high, rather dense cover in a plantation. The majority of countries that produce coffee today regularly practice open-air coffee growing. It is well known that coffee trees with a high potential for productivity can produce large yields when they are grown extensively in the open (Coste, 1992). Although coffee is reputed to be a shade loving plant with a higher quantum utilization efficiency for photosynthesis, excessive shading or light interception by the upper two to three canopy strata of various tree species would reduce growth and productivity of the crop as the plant spends a large portion of its photosynthetic activities for maintenance (Yacob, 1993).The crop's leaves won't have enough reaction centers to hold the light energy and transform it into biochemical energy, on the other hand, if the light intensity is too high. Due to the coffee trees' extreme photorespiration, the majority of their carbs are finally used up. As a result, the trees can experience significant dieback. Additional issues in unshaded coffee orchards include high evapotranspiration, severe drought stress, and death of actively growing shoot sections, seasonal leaf crinkling, frost damage, and eventual output decline (Wrigley, 1988).

Key resource shortages, like a lack of light, can threaten development and survival, whereas abundant sunshine exposes plants to heat, desiccation, and excessive irradiance (Valladares and Niinemets, 2008). Plants have developed a number of well-known biochemical, physiological, and structural changes at the leaf and whole plant levels to deal with these challenges, allowing them to adapt to a specific set of lighting conditions (Lusk *et al.*, 2008; Walters, 2005). The morphological, physiological, and ecological underpinnings of tolerance to extremes, such as tolerance to either sun or shade, have been the subject of several studies on plant light responses. These reactions have frequently been investigated by contrasting plants grown exclusively in high light with those grown in a constant amount of shade (using netting with variable degrees of light penetration) or by contrasting plants grown in gaps with those grown in the forest's understory. Significantly fewer researches, particularly in tropical tree crops, have examined trends in growth and physiological reactions to temporal scales of diurnal fluctuations in light availability. African rainforests' understory is where coffee originated (DaMatta *et al.*, 2007).As a result, its physiology is consistent with a species that has evolved to shade, as evidenced by the slower rates of photosynthesis in full light leaves (7 mol m$^{-2}$ s$^{-1}$) compared to shaded leaves (14 mol m$^{-2}$ s$^{-1}$) (Wintgens, 2004).

2.1.3. Wind

The growth and yield of Arabica coffee may be impacted by wind in several ways. The trees may suffer physical harm as a result of strong winds. The orthotropic and plagiotropic branches may experience a decrease in leaf area and internode length under wind stress (DaMatta *et al.*, 2007). Furthermore, it worsens the loss of budding flowers and fruits (DaMatta and Ramalho, 2006) and severely harms leaves and buds. According to Demel Teketay (1999), cold wind exacerbates the effects of low temperatures, increasing the impact of illnesses referred to as "hot-and-cold" ailments. Water stress in the trees is a result of hot wind increasing evapotranspiration. The effect is substantially more evident in light, highly permeable soils with low retentive capacities where soil-water stocks have been seasonally reduced or depleted (Taye Kufa, 2006).

Desiccation causes wind stress on coffee, which alters the beverage's physiology (Carvajal, 1984).When subjected to artificial wind at speeds as high as 3.0 m s$^{-1}$, Caramori *et al.* (1986) discovered that arabica coffee seedlings engaged in physiological actions that reduced plant height and leaf area. Additionally, the authors noticed a decrease in the internode length of the orthotropic and plagiotropic branches, which may have been caused by a water status impairment brought on by increased transpiration. The connection between the canopy and the surrounding atmosphere should be greatly increased by moderate winds accompanied with lower leaf area. This will raise the conductance of the boundary layer and ultimately enable high transpiration rates. Results by Gutierrez *et al.*, (1994) with field grown adult arabica trees indicate that the apparent stomatal response to wind is mediated by increases in boundary layer conductance, resulting in increased Ds with increasing wind speed. Accordingly, at least a portion of the apparent reactions of coffee to wind can be attributed to stomatal responses to Ds, which are correlated with RH (Gutierrez *et al.*, 1994).As Ds rises with increased wind speed, transpiration would thus decrease as a result of stomatal closure in response to rising evaporative demand.

Crop output is often lower in coffee plantations that are susceptible to strong wind shears and advection. Wind stress can cause significant damage to leaves and buds, as well as increase the shedding of developing flowers and fruits (Matiello *et al.*, 2002). It can also cause a reduction in the leaf area and internode length of the orthotropic and plagiotropic branches (Caramori et al., 1986).The evapotranspiration of crops is increased by hot breezes, which also increases the plants' need for rainfall (or irrigation).It is advised to plant windbreaks or shelter trees where high winds are common since both may enhance crop production. Coffee may be significantly impacted by wind as well. The need for more rainfall (or irrigation) for the trees increases as a result of hot winds increasing evapotranspiration. In addition to worsening the shedding of developing flowers and fruits, strong winds have the potential to seriously harm leaves and buds (Camargo, 1985; Matiello *et al.*, 2002). Wind-breaks must be installed in areas where severe winds are common.

## 2.1.4. Humidity

The coffee tree's vegetative development is greatly affected by humidity. It controls how much water or moisture is lost through evapotranspiration. Water loss decreases when it is high and vice versa. It is crucial, especially during the dry season when high humidity reduces stress on coffee trees and lengthens the time without rain during which the plants may live unharmed. Arabica coffee demands an environment that is less humid, similar to the highlands of Ethiopia (Ramalho, 2006).

The vegetation of the coffee tree is significantly influenced by air humidity. Robusta does well in conditions of high air humidity that are nearly at saturation. It also thrives in less humid environments as long as the dry season is brief. Contrarily, Arabica coffee needs a less humid environment, similar to that found in the Ethiopian highlands (Coste, 1992).

The capacity of the plant to withstand relatively extended periods of soil dryness associated with high atmospheric evaporative demand is significantly impacted by the direct, quick response of stomata to variations in RH. The coffee tree would benefit from such behavior because it would maximize water usage efficiency when soil water supply declines. The sensitivity of stomata to dry air, on the other hand, might be detrimental in conditions of no limiting soil water or temporary water deficiency. In these conditions, increasing agricultural output is more important than increasing water usage efficiency (DaMatta, 2003).

2.1.5. Rainfall

In terms of the ability of climate to alter coffee plant physiology, rainfall ranks second only to ambient temperature (Taye Kufa, 2006). Long term droughts are also difficult for coffee to withstand. According to DaMatta and Ramalho (2006) and DaMatta et al., (2007), the soil's ability to retain moisture, the amount of cloud cover, and cultivation techniques all affect the amount of rainfall that coffee plants need. (Taye Kufa, 2006) Arabica coffee requires an ideal total rainfall range of 1500-1800mm that is uniformly spread during the growth period of 8 to 9 months. Arabicacoffeerequiresabriefdryperiodthatlasts24 months and corresponds to the quiescent development phase in order to blossom (DaMatta and Ramalho,2006).Young wood hardens and flower buds form during the 2-4 drier months when development slows. Carr (2001) suggests that uneven harvests and low yields are frequently caused by the year-round intensity of rainfall (without dry seasons).

Water is equally crucial for the development and growth of plants. Water not only acts as the cellular solvent, enabling chemical processes, but it also encourages cellular volume expansion, completing the carbonic skeleton created as a result of photosynthesis. In addition, water that enters the plant absorbs mineral nutrients and lowers foliar temperature as it evaporates via the stomata. We may use the fact that a maize plant weighing 800 g, during the panicle emission stage, contains around 700 g of water as an example to get a better sense of the amount of water that a plant requires (Boyer, 1995).The volume that transpiration removes from the plant while passing through it is precisely the same volume that the plant uses to transport nutrients to its aerial parts, fix carbon during photosynthesis, and produce an increase in plant mass.

When examining the impact of water on the commencement and differentiation of coffee flower buds, conflicting findings are also frequently seen. These processes are sometimes seen during periods of more humidity, while in other situations, water stress may appear to stimulate them. However, at the period of floral induction, extreme stress might undoubtedly reduce the quantity of inflorescences (Drinnan and Menzel, 1994). As initiation and differentiation of floral buds in coffee are observed at various times of the year, even during the colder seasons, we may thus draw the conclusion that there is no well-defined cycle for these flowers. However, given that there is a concentration of differentiation events during certain times of the year, there appears to be synchronization. By the beginning of March, virtually all of the buds in Campinas, Brazil, had already begun to differentiate, according to an analysis of the local weather (Majerovicz and Söndahl, 2005).This is crucial because 80% of the flowers were connected to the two oldest buds those that had originally differentiated which were between four and five floral buds per leaf axis.

Water is one of the most restricting elements that coffee plants must contend with during their growth (Vicente MR *et al.*, 2017), and it has a direct impact on processes including the timing and frequency of blooming events .Coffee plantations near the equator experience a larger number of flowering events due to a sufficient annual rainfall distribution, but places farther away from this zone experience prolonged dry spells and only experience one major flowering event (Carr MKV., 2001).Only buds at the G4 stage appear to be able to respond to this stimulus, possibly because only these buds have a well-defined vascular cylinder, containing secondary xylem, present on their pedicels. During the dry period, coffee flower buds are thought to become sensitive to respond to the stimulus for regrowth.

Coffee plants' internal, Eco physiological and climatic elements, such as soil moisture, temperature, and climate, all affect their growth and development (Cambrony, 1992). Rainfall is second in significance to ambient temperature as a climatic limiting factor. Water is necessary for the synthesis of carbohydrates, for keeping protoplasm hydrated, and for transporting nutrients and carbs throughout the body. Through its impact on the activity of soil microorganisms, soil moisture level also indirectly influences plant development. The activity of organisms responsible for converting nutrients into forms that are accessible to plants is impeded at extremely low or high moisture levels (Brady and Weil, 2002).

In plants, a lack of water results in a reduction in cell turgor that reaches its greatest value during drought stress, which disturbs the physiological structure of the plant and lowers coffee output

(DaMatta *et al.*, 2006).For coffee plants to produce a good yield, there must be enough water available. Heavy rains and an abundance of moisture have caused several farms to have black rot episodes, bean droppings, and a threat to ripe cherries, which turn a dark shade and some of them shatter. Coffee plants cannot endure prolonged droughts or water logging. When under drought stress, growth is halted; it resumes when the stress is lifted. Additionally, growth, production, and quality are all greatly impacted by drought. The process of losing turgor is likely the one that is most sensitive to drought stress since it impacts the pace of cell growth and final cell size. This is the initial consequence of the stress. As a result, the rate of growth, stem lengthening, leaf expansion, and stomatal aperture are all reduced (Hale and Orcutt, 1987).Cell division and reproductive development are also impacted by severe dehydration. According to Kramer (1983), the impact might vary from fruit or seed maturation and germination to the commencement and development of floral buds. According to Mulualem's research from 1997, water shortages prevent native tree species from growing their cells and organs. Shoot growth is substantially slower, shoot elongation frequently stops (reduces) during noon, and the lowest portions of the stem shrink in seedlings under severe drought stress. Additionally, during water shortages, leaf area decreases, reducing the quantity of photosynthetic energy available for development. Coffee roots won't grow close to permanent water levels and can't endure wet circumstances. Waterlogged soils contain roots that lack root hairs and have bloated, abnormal looking roots (Wrigley, 1988).To promote healthy root and shoot growth in coffee seedlings, ideal soil moisture levels and low to medium soil bulk densities are required (Taye *et al.*, 2002b).

The ideal total yearly rainfall for Arabica coffee is 1,500 to 1,800 mm, equally spaced out during the crop's 8 to 9-month growth season. In terms of soil moisture availability, the ideal coffee-growing agro-ecological zones of Ethiopia range in height (800-3,200 m a.s.l.), temperature (11-26°C), rainfall (500-2,200 mm), and growing season duration (181-300 days) (EARO, 2002). With sufficient moisture, coffee can also be cultivated in warm lowland locations, but it might not be as successful. The minimal amount of yearly precipitation is typically 1,000 mm, spread out throughout the course of the growing season. Coffee cultivation would otherwise require more irrigation. In light of these requirements and the crucial growth stages of a coffee plant, it is crucial to take into account both the quantity and distribution pattern of rainfall.

## 2.2. Soil Factors

Coffee will grow well in soil made from a variety of sources. In reality, it thrives in soils of volcanic origin with a variety of features, including clay-siliceous soils found in granite, as well as alluvial soils (Taye Kufa, 2006).The most critical aspect of soil physiology for coffee plant physiology is excellent drainage. This is due to the fact that the plant cannot tolerate damp soil, which would significantly lower productivity and, if left unchecked, destroy coffee plants (Taye Kufa, 2006; Demel Teketay, 1999).The other two characteristics to take into account are water capacity and depth. Deep soils support root multiplication by supplying a bigger volume of soil which contains more water and nutrients around the coffee plants, while increased water capacity helps to sustain evapotranspiration during dry season (Demel Teketay, 1999).Deep soils are especially important in regions with lengthy dry seasons and little rainfall. Deep soil can support the growth of Arabica coffee. The pH and amount of nutrients present in the soil are additional crucial factors for the growth and production of coffee. Although it may function on slightly acidic soils, the ideal pH range is 5.3 to 6.5 (Taye Kufa, 2006; Charrier and Berthaud, 1985). On soils with a pH of 7.0, however, there are also coffee farms that are quite productive. Very favorable environments include those that are very organically enriched as well as those that are abundant in phosphorus, which is crucial for leaf initiation and shoot growth (Taye Kufa, 2006).

Root dispersion is significantly impacted by soil type. Root penetration and lateral root development are typically hampered by heavier or more compacted soils. The impact of water on the dispersion of roots, however, is not well understood. In a 20 year research conducted in Kenya, the impact of irrigation and mulching (soil covering) using banana peel scraps on the spread of Arabica coffee roots was assessed (Bull 1963).The growth of major (> 5 mm in diameter) and secondary roots in the lower layers is likewise reduced by irrigation, with the main root's deep penetration shrinking (nearly 0.5 m smaller, 20% of the greatest length recorded). However, there was an increase in the lateral length of secondary roots (Emerson Alves da S. and Paulo M., 2008)

The soil needs for coffee don't seem to be very specific. Actually, it works just as well in the clay siliceous soils of granite as it does in volcanic soils with a variety of features or even in alluvial soils. Trees can be killed if waterlogging persists since it significantly lowers production. Therefore, the soil's depth and texture are crucial components. The root system of the coffee tree may grow very far. A depth of more than 150 cm is needed for it to be effective. Due to this quality, it is able to utilize a vast area of land and hence make up for a relative lack of fertility. The organic matter level of the soil in highly appropriate locations was high (>3%). A somewhat acidic soil is preferable in terms of PH. The ideal pH range is between 5.3 and 6.5. On almost neutral (pH 7.0) soils, there are, nevertheless, also very productive coffee farms

(Clifford, 1985; Coste, 1992; Paulos, 1994).The root development is most significantly influenced by nitrogen as a single element. Shoots, however, are unable to use nitrate because they lack nitrate reductase. Initiation of new leaves and shoot development depend on phosphorus. Therefore, phosphatic fertilizers should be used to promote quicker development of suckers when shoot growth is more necessary than root growth.

According to Demel Teketay (1999), altitude has a direct impact on climate, which changes how coffee functions physiologically. Altitude has a considerable impact on temperature since it causes a drop in temperature of roughly 6 ·C for every 1000 m in elevation (Demel Teketay, 1999). Arabica coffee is better cultivated at higher elevations and is frequently farmed in mountainous, steep terrain (Wrigley, 1988).

It is an upland species that is endemic to the tropical forests of southwest Ethiopia and grows as an understory tree at elevations between 1600 and 2800 meters (DaMatta & Ramalho, 2006). The optimal circumstances for the growth of *Arabica coffee* are found on the equator at around 1525-1830m above sea level (Demel Teketay, 1999), while the height for coffee cultivation varies from nation to country. For instance, the ideal altitudes for climbing Mount Kilimanjaro in Tanzania, Kenya, and Mexico are 1370–1680 m, 1590–1770 m, and 920 m, respectively. According to Willson (1985), referenced in Demel Teketay (1999), Arabica coffee is a highland crop that grows between 1000 and 2000 meters above sea level in tropical regions. Coffee arabica may be produced at lower elevations farther from the equator up until it is constrained by frequent and protracted frost (Taye Kufa, 2006; Demel Teketay, 1999).

Other elements, such as geographic location (latitude, longitude), soil, etc., greatly influenced how big of an impact elevation had. The main variable that varies with elevation is temperature, which also changes with latitude and longitude (Austin *et al.*, 1984). Elevation is an indirect environmental gradient (it has no direct impact on plant physiology).Up to a height of about 1600 m above sea level, a rise in altitude caused an increase in bean size and soil organic matter, but there was no further substantial increase (i.e., a hump shaped relationship rather than a monotonic relationship) (Abebe *et al.*, 2020). This may be linked to a decline in the mineralization and decomposition of soil organic matter, which occurs as a result of the drop in temperature with elevation. According to a research by Alpizar and Bertrand (2004), the proportion of large sized beans increases with elevation; this relationship was shown up to a height of 1400 m above sea level before it began to fall. Higher carbon-to-nutrient ratios, such as C: N, C: P, etc., are indicative of decreased nutritional availability (Wilcke *et al.*, 2003,

2008).At higher elevations in the current investigation than at lower elevations, the C: P and C: N ratios were considerably greater (data not shown). Thus, decreasing nutrient availability at higher elevations might be the probable reason for decreasing bean size after an elevation gradient of 1600 m.

As discussed by (Abebe *et al.*, 2020) coffee physiological activities was significantly varied across the elevation ranges, being altered in the elevation range 1300-1600 m above sea level (asl).It was also interesting to see that altitude had a negative impact on fruit load and productive nodes, or that it had no influence on productive branches per stem and cherry weight. Alejandra S. (2021) anticipated a stronger positive altitudinal effect with higher yield component values at mid- and high altitude than at lower altitude due to unfavorable conditions, particularly high temperature and a high vapor pressure deficit that result in unfavorable growing conditions (Sarmiento-Soler *et al.*, 2019).

# 3. CONCLUSION

The tropics are home to more than 70 species of coffee (genus *Coffee*).Africa is the source of all domesticated species. Coffee arabica (Arabica, 64% of global output) and Coffea canephora (var. Robusta, 35%) are today's two most significant coffee species economically. Given that it is grown on around 10.3 million hectares globally and provides more than 25 million households with their sole source of income, coffee is one of the most significant agricultural products on the global market. It is one of the most important cash crops in underdeveloped countries and is grown and sold by more than 60 countries.

The plantation crop coffee is well suited to the various Eco physiological conditions of the tropics. Above 1000 m height along the equator, C. arabica is suited to colder temperatures; at higher latitudes, the temperature is a little lower. As a result, the crop thrives on tropical highlands. The quality of C. canephora increases with height, however it likes a hotter temperature and is better adapted to lowlands (below 900 m altitude).Because of its shorter root structure, it needs a protracted wet season. It can handle excessive soil moisture, but it needs a brief dry season for significant flowering.

The tropical forests of Africa are all of the coffee species' native habitats. Arabica and Robusta varieties of coffee are highly adapted plantation crops to the various eco-physiological conditions of tropical highlands. In finding the ideal or nearly ideal edapho climatic conditions for cultivation, altitude and latitude must be taken into account together with temperature, rainfall and water supply, soil, slope, and aspect. At the equator, Arabica coffee requires elevations of between 1,200 and 2,200 m above sea level, while the ideal height might be lower at higher latitudes. As long as the temperature requirements (15 to 30° C) are met, Arabica coffee is grown between 25°N and 24°S, ranging from more inclined subtropical areas to tropical regions with greater altitudes. Due to its origin in central Africa, Robusta coffee thrives in tropical locations with lower elevations with hot and humid weather. The physiology of the coffee plant is also influenced by other environmental parameters, such as soil type, soil PH, texture and structure, soil water holding capacity, and altitude, through impacting root activity, flower initiation, bud development, and metabolic activity

# 4. REFERENCES

Abebe Yadessa, Juergen Burkhardt, Endashaw Bekele, Kitessa Hundera and Heiner Gold Bach (2020) .The major factors influencing coffee quality in Ethiopia: The case of wild Arabica coffee (*Coffee arabica* L.) from its natural habitat of southwest and southeast afromontane rainforests. African Journal of Plant Science, Vol. 14(6), pp. 213-230, June 2020

Alejandra sarmiento Soler (2021) coffee productivity and water use in open vs shaded systems along an altitudinal gradient at mt. Oregon, Uganda.

Araujo, W. L. Dias, P.C. Moraes, G.A.B.K., Celin, E.F., Cunha, R.L., Barros, R.S. and DaMatta, F. M. (2008). Limitations to photosynthesis in coffee leaves from different canopy positions. *J. plant Physiol. and Biochem.* 46: 884-890

Barbosa J.N., Borém F.M., Cirillo M.Â., Malta M.R., Alvarenga A.A., Alves H.M.R. Coffee quality and its interactions with environmental factors in Minas Gerais, Brazil. *J. Agric. Sci.* 2012; 4(5):181. [Google Scholar]

Barros, R.S., Maestri, M., Rena, A.B., (1999) Physiology of growth and production of the coffee tree— a review. *J. Coffee Res.* 27, 1–54.

Belete Y., Bayetta B., Chemeda F. Stability analysis of bean yields of Arabica coffee genotypes across different environments. *Green. J. Plant Breed. Crop Scie.* 2014; 2:18–26. [Google Scholar]

Bertrand B, Bardil A, Baraille H, Dussert S, Doulbeau S, (2015) The greater phenotypic homeostasis of the allopolyploid *Coffee arabica* improved the transcriptional homeostasis over that of both diploid parents. Plant & Cell Physiology 56:2035–51

Bunn C, Läderach P, Ovalle Rivera O, Kirschke D. (2015) A bitter cup: climate change profile of global production of Arabica and Robusta coffee. *Climatic Change* 129:89–101

Charbonnier, F.; Roupsard, O.; le Maire, G.; Guillemot, J.; Casanoves, F.;Lacointe, A.; Vaast, P.; Allinne, C.; Audebert, L.; Cambou, A.; Clément-Vidal, A.; Defrenet, E.; Duursma, R.A.; Jarri, L.; Jourdan, C.; Khac, E.; Leandro, P.; Medlyn, B.E.; Saint-Andre, L.; Thaler, P.; Van Den Meersche, K.; Barquero Aguilar, A.; Lehner,P.; Dreyer, E.( 2017) Increased light-use efficiency sustains net primary productivity of shaded coffee plants in agroforestry system. *Plant Cell Environ.* 40, 1592–1608.

Cheserek J.J., Gichimu B.M. (2012) Drought and heat tolerance in coffee: a review. *Int. Res. J. Agricult. Sci. Sci.* 2012; 2(12):498–501. [Google Scholar]

CLOWES, M. ST. J., NICOLL, W. D. and SHELLEY, R. S. (Eds). (1989). Coffee manual for Malawi 1989.Tea and Coffee Research Foundation of Central Africa, Malawi.

Da Matta, F. M. (2004). Ecophysiological constraints on the production of shaded and unshaded coffee: a review. *Field Crops Research, 86*(2-3), 99-114. doi:10.1016/j.fcr.2003.09.001

Da Matta, F.M. and Ramalho, J.D.C. (2006). Impacts of drought and temperature stress on coffee physiology and production: a review. Braz.J.Plant Physiol. 18:55-81

DaMatta FM, Rahn E, Läderach P, Ghini R, Ramalho JC. (2019) Why could the coffee crop endure climate change and global warming to a greater extent than previously estimated? Climatic Change 152(1):167–17

DaMatta, F. M. (2018) Coffee Tree Growth and Environmental Acclimation. In Achieving Sustainable Cultivation of Coffee; Lashermes P., Ed., in press; Burleigh Dodds Science: Cambridge, UK; pp. 19-48.

DaMatta, F. M., Rahn, E., Läderach, P., Ghini, R., & Ramalho, J. C. (2018). Why could the coffee crop endure climate change and global warming to a greater extent than previously estimated? Climatic Change, 152(1), 167-178.doi:10.1007/s10584-018-2346-4

De Oliveira RR, Ribeiro THC, Cardon CH, Fedenia L, Maia VA, (2020) Elevated temperatures impose transcriptional constraints and elicit intraspecific differences between coffee genotypes. Frontiers in Plant Science 11:1113

Gole Tadesse W., Itana A., Tsegaye B., Senbeta F.(2015) *Coffee: Ethiopia's Gift to the World the traditional production systems as living examples of crop domestication, and sustainable production and an assessment of different certification schemes. Environment and Forest Coffee Forum.* Environment and Coffee Forest Forum; Addis Abeba, Ethiopia: 2015. [Google Scholar]

ITC (International Trade Center) third ed.( 2011). The Coffee Exporter's Guide. [Google Scholar]

Krishnan, S.; Ranker, T. A.; Davis, A. P.; Rakotomalala, J. J.( 2013) A 577 assessment of the genetic integrity of ex situ germplasm collections of three endangered species of *Coffee* from Madagascar: implications for the management of field germplasm collections. *Genet. Resour. Crop. Evol. 60*, 1021–1036.

Laderach P., Oberthur T., Cook S., Iza M.E., Pohlan J.A., Fisher M., Lechuga R.R. (2011) Systematic agronomic farm management for improved coffee quality. *Field Crop. Res.* 2011; 120:321–329. [Google Scholar]

Martins MQ, Rodrigues WP, Fortunato AS, Leitão AE, Rodrigues AP, (2016) Protective response mechanisms to heat stress in interaction with high [CO2] conditions in *Coffee spp. Frontier in Plant Science* 7:947

Martins, S. V. C.; Galmés, J.; Cavatte, P.C.; Pereira, L. F.; Ventrella, M. C.; DaMatta, F.M. (2014)Understanding the low photosynthetic rates of sun and shade coffee leaves: Bridging the gap on the relative roles of hydraulic, diffusive and biochemical constraints to photosynthesis. PLoS One, 9, e95571.

Mayer JLS, Carmello-Guerreiro SM, Mazzafera P. (2013) A functional role for the collectors of coffee flowers. *AoB PLANTS* 5:plt029

Poorter H.Sack.L. (2012) Pitfalls and Possibilities in the analysis of biomass allocation patterns in plants. Frontiers in plants science doi:10.3389/fpls.2012.00259.

Sarmiento-Soler, A., Vaast, P., Hoffmann, M. P., Rötter, R. P., Jassogne, L., van Asten, P. J. A., & Graefe, S. (2019). Water use of *Coffee arabica* in an open versus shaded system as under smallholder's farm conditions in Eastern Uga Agricultural and Forest Meteorology, 266-267, 231-242. doi:10.1016/j.agrformet.2018.12.006

Sridevi V., Giridhar P. (2013) Influence of altitude variation on trigonelline content during ontogeny of *Coffea canephora* fruit. *J. Food Stud.* 2013; 2:62–74. [Google Scholar

Steiman S.(2013) Why does coffee taste that why? Notes from the field. In: Thurston R.W., Morris J., Steiman S. Coffee, editors. *A Comprehensive Guide to the Bean the Beverage and the Industry.* Rowman and Littlefield; USA: 2013. [Google Scholar]

# YOUR KNOWLEDGE HAS VALUE

- We will publish your bachelor's and
  master's thesis, essays and papers

- Your own eBook and book -
  sold worldwide in all relevant shops

- Earn money with each sale

Upload your text at www.GRIN.com
and publish for free